Farmer

Margaret Hudson

Contents

Heinemann

Where in the world?

Farmers grow the food we eat. All over the world there are farmers working to grow food for themselves and for other people.

We are going to visit four farmers.

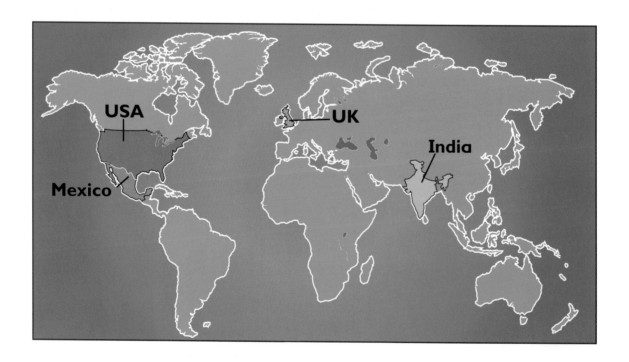

Jim is a **dairy farmer**. His farm is in Herefordshire, in the United Kingdom (UK).

Bapa is a farmer in Kesharpur **village**, in Orissa, India.

Jack is a peanut farmer in Geneva, Alabama, in the United States of America (USA).

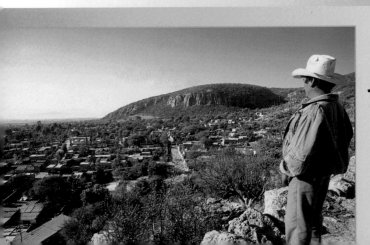

Juan is a farmer in La Trinidad, Mexico, in Latin America.

Farming

Different farmers work in different ways.

Jim has about 70 cows. He sells almost all the milk his cows make. He has to keep a record of how much milk he sells.

Bapa and the other farmers in Kesharpur **village** grow lots of different **crops**. They eat most of what they grow.

Jack uses machinery on his farm.
The tractor **ploughs** a lot of rows
at once.

Juan does not have a tractor.
He uses horses to pull his plough.
He can only plough one row at a time.

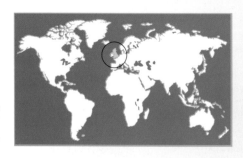

United Kingdom

Jim and Sheila Hitchon have three children, Will, Sarah and Shirley. They live in a farmhouse in the middle of their land. They have to drive to the nearest town to shop.

The family are well known for their
herd of **Clenchers** cows. But they
grow some **crops** of maize and wheat
too. Here Jim is **ploughing** the soil.
The land is quite hilly, so it can be
hard work.

Will works on the farm all the time.
Sarah and Shirley go to school, but
sometimes help. Here they are milking
one of the cows. They have fixed the
milking machine to the cow's **udder**.
It pumps milk out. The machine can
milk several cows at the same time.

The family eat their main meal in the middle of the day. Sometimes friends come to eat too. They are eating lamb, potatoes and peas. After that they will eat **trifle**.

India

Udayanatha Khatei, usually called Bapa, lives in one of these houses in Kesharpur **village**. There is forest all around. The villagers grow wheat, **sugar cane**, vegetables and rice in the fields around the village. They keep goats, pigs, **buffalo**, **cattle** and chickens.

Bapa and the other villagers are **harvesting** the wheat. They do all the work by hand. They work together until all the fields are harvested. They take the **crops** back to the village in carts.

It is important that the villagers protect the land and keep the forest healthy. They do this by working together to clear fields and plant trees.

While the men work in the fields the women and children fetch water, look after the animals and prepare the food. The children also go to school. The **village** has its own schools.

The family eat their main meal in the
middle of the day. At busy times of the
year, like **harvest** time, the children
will bring a meal of rice out to the
fields and eat there. Bapa and the
others will be too busy to go back
to the village to eat.

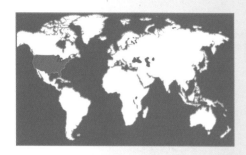

USA

Jack and Ann Ausley and their daughter Beth live in a house in the middle of their farmland. Their daughter Missy lives nearby. They have to drive to the nearest shops and school.

Jack grows and sells peanuts. They are his main **crop**. But the Ausleys grow cotton and keep **cattle** too – 135 cows and 6 bulls.

Jack is feeding the animals **peanut hay**. The cows graze in the fields too, but when there is no fresh grass Jack gives them hay. Jack sells the **calves** when they are about a year old.

Ann helps Jack to keep a record of all the things they buy and sell. She also cooks the meals. She spends a lot of time bottling and freezing home-grown vegetables and fruit, so they can eat them all year round.

A farm worker, Willy Copes, helps Jack with the heavy work on the farm.

The family eat their main meal in the
evening. Missy and her husband Steve
often come to dinner.

They are eating fried chicken,
potatoes, corn and corn bread. After,
they will eat fresh fruit and pecan nut
pie. The vegetables and fruit are
home-grown.

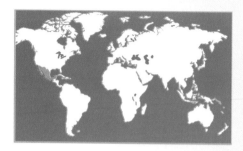

Mexico

Juan Ugalde, his wife Guillermina and their children live in the **village** of La Trinidad. They grow some vegetables in the garden by the house. But, like the rest of the villagers, Juan farms the fields around the village. He walks to work each day.

18

The children help when they are not in
school. Their son Edgar, and one of his
cousins, are taking the sheep to the
fields outside the village. The children
water the garden, help in the fields and
help Guillermina in the house.

When he is not working in the fields, Juan does other jobs around the **village**. All the villagers share the jobs that have to be done to keep things in good repair.

Here Juan and his cousin are repairing a stone wall.

The family eat their main meal in the middle of the day. They are eating chilli **rellenos** and **tortillas**.

Edgar, Marisol, Manuel, Rosio and Miguel-Angel are eating. Guillermina is serving the food.

Factfile

India

Population: 943 million

Capital city: New Delhi

United Kingdom (UK)

Population: 58 million

Capital city: London

Mexico

Population: 93 million

Capital city: Mexico City

United States of America (USA)

Population: 264 million

Capital city: Washington DC

Digging deeper

1 How is the landscape different between the UK and Mexico? Look at pages 3 and 5.

2 Look at the houses in India on page 10. How are they different to your home?

3 What do you think the weather is like in Alabama, USA? Why do you think this?

Glossary

buffalo   a kind of ox, related to cattle

calves   young cows

cattle   male and female cows

Clenchers   a type of Friesian cow

crops   plants grown by farmers

dairy farmer   a farmer who keeps cows for their milk, which is then sold

harvest(ing)   collecting ripe crops

herd   lots of animals of the same type all together

peanut hay   the stalks of harvested peanuts

ploughing   turning over and breaking up the soil ready for planting

tortillas   flat pancakes made from flour or ground maize

trifle   a sweet dish usually made with layers of fruit, custard and cream

rellenos   peppers stuffed with cheese

sugar cane   one of the plants that sugar is made from

udder   the part of a cow that gives milk

village   small groups of people living in one place

Index

First published in Great Britain by Heinemann Library
Halley Court, Jordan Hill, Oxford OX2 8EJ
a division of Reed Educational and Professional Publishing Ltd

OXFORD FLORENCE PRAGUE MADRID ATHENS MELBOURNE
AUCKLAND KUALA LUMPUR SINGAPORE TOKYO IBADAN
NAIROBI KAMPALA JOHANNESBURG GABORONE
PORTSMOUTH NH CHICAGO MEXICO CITY SAO PAULO

© Reed Educational and Professional Publishing Ltd 1996

Designed by John Walker

Illustrations by Oxford Illustrators and Visual Image

Printed in Malaysia

00 99 98 97 96
10 9 8 7 6 5 4 3 2 1

ISBN 0 431 06339 7

British Library Cataloguing in Publication Data

Hudson, Margaret
Farmer
1. Farmers – Juvenile literature
I. Title
630.9'2

Acknowledgements

The Publishers would like to thank the following for permission to
reproduce photographs:

Steve Benbow: pp. 3, 5,14-17;

Chris Honeywell: pp. 3, 4, 6-9;

Rajendra Shaw/Oxfam: pp. 3, 4, 10-13;

Sean Sprague: pp. 1, 3, 5, 18-21

Cover photograph reproduced with permission of
Rajendra Shaw/Oxfam

Our thanks to Clare Boast for her comments in the preparation
of this book.

Every effort has been made to contact copyright holders of any
material reproduced in this book. Any omissions will be rectified
in subsequent printings if notice is given to the Publisher.

Oxfam believes that all people have basic rights: to earn a living,
to have food, shelter, health care and education. There are nine
Oxfam organizations around the world - they work with poor
people in over 70 countries. Oxfam provides relief in
emergencies, and gives long term support to people who are
working to make life better for themselves and their families.

Oxfam (UK and Ireland) produces a catalogue of resources for
schools and young people. For a copy contact Oxfam, 274
Banbury Road, Oxford, OX2 7DZ (tel. 01865 311311) or contact
your national Oxfam office.

Oxfam UK and Ireland is a Registered Charity number 202918.
Oxfam UK and Ireland is a member of Oxfam International.